THE MARVELOUS CONTRAPTION

TRACY LEWIS

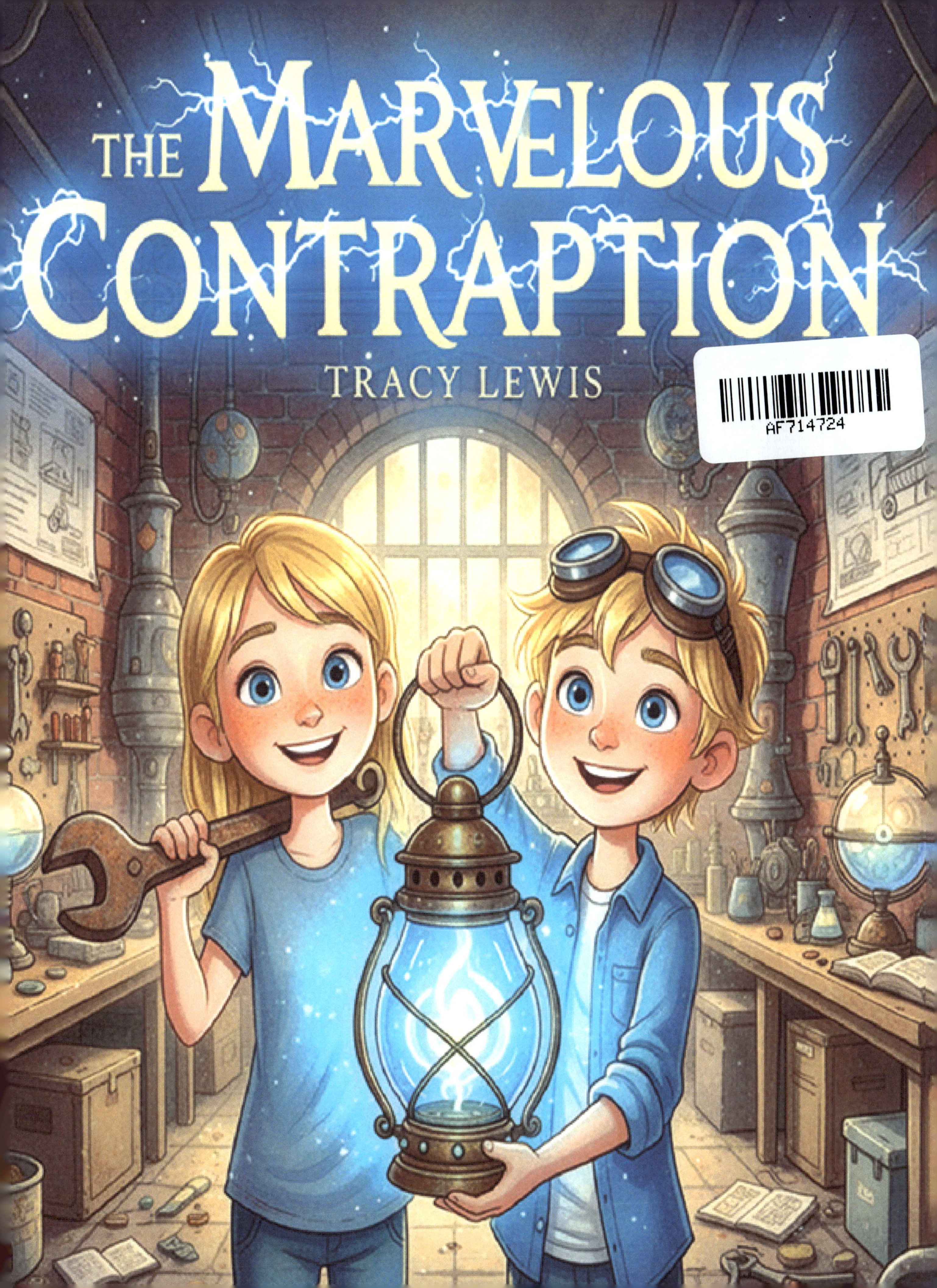

Table of Content

Chapter 1:

The Boy Who Saw Gears in Clouds

Wilson Tinker didn't just see the world - he reverse-engineered it. At ten years old, his bright blue eyes were perpetually wide with wonder, dissecting everything from the aerodynamics of falling leaves ("Miniature parachutes!") to the hydraulic hiss of the school bus brakes ("Piston-powered poetry!").

His messy blonde hair seemed charged with static electricity, constantly defying gravity like his ideas.

The Tinker household hummed like a well-oiled (if occasionally sputtering) machine.

Wilson occupied the middle seat in their family's symphony - sandwiched between fifteen-year-old Wesley, whose corner of the living room was a glowing fortress of monitors emitting electronic beep-boop symphonies, and eleven-year-old Lyla Jo, whose fashion sketches bloomed across every flat surface like colorful weeds.

Downstairs, seven-year-old Blair drilled herself on "Assistant Protocols," marching through the kitchen with a toy stethoscope checking the "vital signs" ofappliances.

The garage was Wilson's cathedral of creativity - a glorious, chaotic shrine to "What If?" Shelves sagged under the weight of ambitious prototypes in various states of glorious failure:

- The Automatic Sandwich Maker (now fossilizing a peanut butter-jelly geological layer)
- The Turbo-Dog-Walker 3000 (Jax the Labrador had enthusiastically dragged it into Duck Pond #2)
- The Homework-Doing Robot (which answered every math problem with "01100101" - binary for 'e')

"Why's your head always in the clouds, Willy?" Wesley teased one spaghetti night as Wilson sketched gear ratios on his napkin.

"Because that's where the blueprints are!" Wilson shot back, adding a quick doodle of a pasta-powered engine beside his meatballs.

Lyla Jo sighed dramatically, adjusting her handmade beaded necklace. "Must you turn everything into schematics? Even Mom's lasagna?"
Wilson grinned, holding up a cheese-covered noodle. "This isn't lasagna... it's edible structural engineering!

Wilson's mother, Katy - whose race cars sliced through wind at 200 mph - would step over tangled wires to kiss his flour-dusted forehead. "Every masterpiece starts as a beautiful disaster, inventor-man." His father, Matt, CEO of PetroCo, brought exotic treasures from his commute: "Found this titanium mesh near the refinery fence... looked lonely!"

But Wilson's true compass pointed next door. Nana's kitchen was a sanctuary of cocoa steam and wisdom. Her elegant hands - adorned with rings from Poppy's travels (an emerald from Brazil, a turquoise from Arizona) moved like graceful conductors as she poured drinks. Her magnificent black curls seemed to hold entire constellations of stories, and Wilson often imagined equations hidden in their spirals.

Poppy smelled perpetually of pine sawdust and possibility, his full head of silver hair gleaming like polished chrome under the workshop lights. Their boxer JoJo guarded Wilson's projects

Chapter 2:
The Impossible Idea

The storm arrived like an angry orchestra - thunder booming its bass notes, rain drumming on the roof, wind whistling through the eaves. Wilson pressed his nose against the cold windowpane, watching lightning fork across the bruised-purple sky. Crack-BOOM! A jagged bolt split the darkness, and suddenly Wilson saw it - not destruction, but raw, pulsing energy.

"What if..." he breathed, eyes wide. Blair looked up from coloring her dragon (currently wearing Lyla Jo's hat).
"What if what?"
"What if we could tame it?" Wilson whispered, his mind already racing faster than the raindrops. "Harness it... like..."

Blair dropped her crayon. "Like a lion?"
Wilson shook his head, blonde hair flopping. "Safer! Like catching fireflies... but zappy! We'd bottle the energy!"
Blair gasped as Wilson snatched his sketchbook. His pencil became a blur - coils, conductors, containment fields spilling across pages.

By dawn, "Project SkySpark" had colonized Wilson's walls. Blueprints layered over dinosaur posters; equations framed his window. Wilson moved through the next weeks in a fever dream. He missed soccer practice ("Newton's Laws are more fun applied!"), left homework half-done ("Calculating voltage drop takes priority!"), and murmured to transformers at dinner ("You get me, don't you?").

"Wilson," his mother said gently one evening, finding him asleep curled around a capacitor like a metallic teddy bear. She brushed flour from his cheek - he'd been using the kitchen scale to weigh copper filings. "Even race engines need pit stops, inventor-man."

His father frowned later, inspecting Wilson's singed eyebrows. "Safety protocols, champ. Always."
Lyla Jo stormed into the living room, Jax trailing behind her. "Look!" She pointed at the Labrador's tail tip, glowing an eerie, persistent blue. "Wilson zapped Jax again!"
Jax wagged happily, apparently unconcerned.

Failures piled like discarded pizza boxes:

- Prototype #1: Melted into a modern-art puddle ("Abstract Expressionism!" declared Nana).
- Prototype #2: Gave Mazel a tiny shock. The dachshund yelped, then forgave Wilson for three strips of bacon ("Strategic bribery," noted Poppy).
- Prototype #42: Accidentally turned Blair's ponytail neon green for six hours. ("Ooh!" she'd chirped, admiring herself in the mirror. "Can it be permanent?")

One rainy midnight, after Prototype #68 sputtered and died with a sad pffft, Wilson kicked a coil across the garage. It clanged against Wesley's old bike. "Maybe Wesley's right!" he muttered, blinking back hot tears.

"Maybe it is impossible..."

The garage door creaked open. Poppy stood there, two steaming mugs in hand, rain glistening on his silver hair. "Edison failed a thousand times before he lit the world," he said softly, handing Wilson cocoa.

"But I'm not Edison!" Wilson's voice cracked. Poppy's eyes crinkled. "No... you're Wilson Tinker. And Tinkers?" He tapped Wilson's chest gently. "Persist." Wilson clutched the warm mug, its heat seeping into his cold fingers. Outside, thunder rumbled like God clearing His throat. Wilson closed his eyes. "Help my hands build what You've shown my heart," he prayed into the steamy darkness. "And maybe... keep Jax's tail normal?"

Chapter 3:
Sparks Fly

The numbers flashed behind Wilson's eyelids like stubborn fireflies - 47.8 Hz... 49.2 Hz... - until suddenly, they snapped into perfect alignment. Wilson bolted upright at 3:17 AM, his heart hammering against his pajamas. "RESONANCE FREQUENCY!" he gasped into the darkness. "Like when Nana sings into her water glasses! Or whistles with an Acorn."

He scrambled downstairs, nearly tripping over Mazel who blinked sleepily from her dog bed. In the garage, Wilson yanked the chain of the single dangling bulb. Its sudden click-hum threw long, dancing shadows over the chaos of coiled wires and metal scraps. With trembling hands, he dialed Blair. Three rings later, her groggy voice mumbled, "...Whaaa?"

"RED ALERT!" Wilson whispered fiercely. "SkySpark Phase Three - ASSEMBLE!"

There was a thump, then Blair's suddenly wide-awake voice: "Roger! ETA two minutes!"

True to her word, Blair arrived clutching Mr. Snuggles (her stuffed owl) in one hand and her Official Assistant Toolkit in the other. Her blonde hair stuck up in wild tufts, but her eyes behind fogged goggles were laser-focused. "Status report, Captain!"

"Critical phase!" Wilson handed her a blueprint. "We need crystalline Rod Gamma recalibrated to 48.5 megahertz - STAT!"

Blair saluted, her freckles serious in the yellow light. "Aye!" She rummaged in her toolkit. "Quantum Spanner, incoming!" (It was a slightly bent wrench from Poppy's shed).

As Wilson adjusted the delicate rods - recycled from Nana's old chandelier - he remembered Poppy's voice: "The storm outside's just noise, Willy-boy. But when the storm inside you matches its rhythm? That's when magic happens."

Wilson took a shaky breath. Outside, thunder grumbled like a sleeping giant. Match the rhythm...

"Ready?" Wilson's finger hovered over the salvaged helicopter toggle switch. Blair clutched Mr. Snuggles tight, nodding silently.

Click.

Silence swallowed the garage. Wilson's stomach dropped. Another failure...

Then... a soft, deep thrummmmmm vibrated up through the workbench. Like a cello string plucked deep underwater. The crystalline rods began to glow - first faint amber, then brilliant sapphire.

Pop! Fizz! Tiny bolts of blue lightning - no thicker than thread - danced inside the glass sphere. They swirled and intertwined, casting flickering constellations across Blair's awestruck face, glinting off discarded screwdrivers, painting the ceiling with living constellations.

Blair's whisper trembled. "Is it... alive?"
Wilson stared at the captured storm, tears blurring the brilliant blue. His voice cracked like dry kindling. "It's... it's alive!"

He swept Blair into a spinning hug, their laughter - pure, unfiltered joy - bouncing off toolboxes and echoing into the rain-lashed night. Outside, lightning flashed again, but now it felt like the sky was cheering.

Chapter 4:

The Town Square Revelation

Maple Creek hadn't witnessed such commotion since Old Man Henderson's prize-winning pumpkin escaped its crate and rolled through Mrs. Gable's prize petunias.

By noon, the town square thrummed with energy. Lawn chairs sprouted like colorful mushrooms; kids darted between legs clutching sparklers (despite the daylight); even Mrs. Pettigrew's normally aloof Persian cat watched wide-eyed from a bakery windowsill.

Mayor Thompson adjusted his best paisley tie beside the velvet-draped invention, beaming like a lighthouse spotting safe harbor. Wilson fidgeted, suddenly hyper-aware of the grease stain on his left knee. Why didn't I wear Lyla Jo's "fancy" overalls? His palms felt slick.

His family formed a proud, slightly chaotic phalanx at the front:

- Mom Katy filmed through proud tears, whispering, “Deep breaths, inventor-man!”
- Dad Matt stood ramrod straight, though his stern CEO expression kept cracking into a grin.
- Blair bounced in her laminated “Official Assistant” badge, clutching Mr. Snuggles the owl under one arm.
- Lyla Jo’s custom sequined “Inventor’s Sister” beret glittered defiantly (“Sequins show solidarity!” she’d declared, though Wilson noticed she stood just outside grease-splatter range).
- Even Wesley lowered his VR headset long enough to shout, “Don’t blow anything up, Zap-Boy!”

Nana clutched Poppy's arm, her rings fracturing the midday sun into tiny rainbows. "Our boy," she breathed, voice thick as honey. Poppy squeezed her hand, his own eyes suspiciously bright.

"People of Maple Creek!" Mayor Thompson's voice boomed through crackling speakers. The crowd hushed, leaning forward as one. "Today, we celebrate not just ingenuity... but grit!" He gestured grandly toward Wilson. "Meet young Wilson Tinker - and his Lightning Lantern!"

Wilson's heart hammered against his ribs. Don't trip. Don't drop it. Please, God... With a flourish, he pulled the velvet cloth. A collective gasp rippled through the square like a summer breeze through wheat.

Inside the glass sphere, harmless bolts of sapphire-blue energy danced and intertwined - less like captured lightning, more like bottled Northern Lights. Ethereal light washed over upturned faces, painting children's cheeks blue and making Mrs. Gable's glasses glow.

"This young man," the Mayor's voice thickened with emotion, "showed us that tenacity turns 'impossible' into 'I'M POSSIBLE!'" He lifted a gleaming golden wrench trophy - its handle shaped like a lightning bolt. "For unwavering persistence!"

Cheers erupted - whistles, thunderous claps, Jax's booming barks from Dad's arms. Wilson's knees wobbled. He leaned into the microphone, its cold metal strangely grounding. "My lantern isn't just about lightning..." His voice, clear and strong, surprised him. "It's about not letting doubts - yours or anybody's - snuff out your spark!" He caught Poppy's encouraging nod.

"God gave us curiosity like a compass..." He glanced at Blair, bouncing on her toes. "...and sometimes, He gives stubborn kids loud ideas!" Laughter rippled warmly. "So follow that compass! Even when it points straight through the thunder!"

LIGHTNING LANTERN FESTIVAL

Later, kneeling by his bed beneath the glow of his yellow star nightlight, Wilson surveyed his blueprint-strewn room. The golden wrench trophy sat beside his dented third-place soccer trophy. Outside, the real lantern glowed softly over Maple Creek.

Thank You, he prayed, clasping his hands tight. For Nana's cocoa courage, Poppy's toolbox wisdom, Blair's Quantum Spanners... and letting loud ideas light up the dark.

Epilogue: The Garage Never Sleeps

The Lightning Lantern now glows in City Hall, powering emergency lights during storms. Wilson still fills the garage with wild ideas ("Anti-Gravity Socks!"). But now, neighborhood kids visit not just to see inventions... but to invent together.One Saturday, Lyla Jo even designed "Inventor-Chic" overalls ("Grease stains are texture!"). Wesley helped code a safety monitor ("No more glowing dogs, okay?"). Blair remained Chief Assistant, though she now demanded a lab coat "for professionalism."

As Poppy watched Wilson teach little Miguel how to solder, he squeezed Nana's hand. "Our boy's building more than gadgets..."

Nana's rebuttal as she wiped a tear. "He's building believers."

Nana watched Wilson adjust Blair's oversized goggles, his laughter mingling with the kids' chatter. Her eyes shone. "He's building believers, Danny. Little torchbearers."

For Wilson Tinker finally understood what Poppy whispered that stormy night: Marvelous contraptions aren't just bolts and blueprints.

They're faith that ignites when skies are darkest, family who steadies your hand and passes the spanner, and the stubborn, joyful noise of staying true to that quiet voice inside - even when the world shouts "Stop!" Because sometimes, if you hold fast to your compass through the thunder... you don't just light a lantern. You become the light others follow to their own impossible dreams.

THE END

www.ingramcontent.com/pod-product-compliance
Lightning Source LLC
Chambersburg PA
CBHW041323210926
51847CB00002B/32
* 9 7 8 1 9 6 8 5 2 5 4 9 1 *